Our Universe

Pluto

by Margaret J. Goldstein

Lerner Publications Company • Minneapolis

Lerner Publications Company
A division of Lerner Publishing Group
241 First Avenue North
Minneapolis, MN 55401 USA

Website address: www.lernerbooks.com

Words in **bold type** are explained in a glossary on page 30.

Library of Congress Cataloging-in-Publication Data

Goldstein, Margaret J.
    Pluto / by Margaret J. Goldstein.
      p.   cm. — (Our universe)
    Includes index.
    Summary: An introduction to Pluto, describing its place in the solar system, its physical characteristics, its movement in space, and other facts about this outer planet.
      ISBN: 0–8225–4656–6 (lib. bdg. : alk. paper)
      1. Pluto (Planet)—Juvenile literature. [1. Pluto (Planet)]
I. Title. II. Series.
QB701 .G65 2003
523.48'2–dc21
                                       2002000430

Manufactured in the United States of America
1 2 3 4 5 6 – JR – 08 07 06 05 04 03

Some planets shine in the night sky.
But you will not see the planet Pluto.
Do you know why?

Pluto is too far away for us to see. It is on the outer edge of the **solar system.** The solar system includes nine planets in all. Earth is part of the solar system.

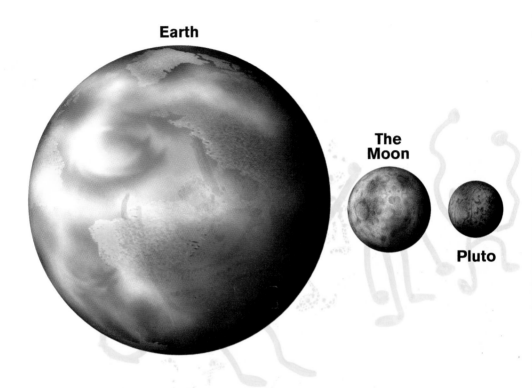

Earth

The
Moon

Pluto

Pluto is the smallest planet in the solar system. It is even smaller than Earth's moon.

All of the planets in the solar system **orbit** the Sun. To orbit the Sun means to travel around it.

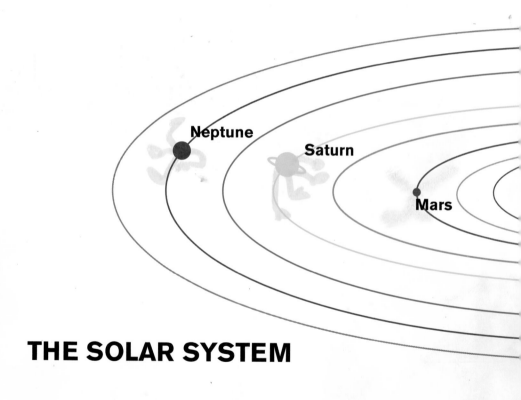

THE SOLAR SYSTEM

Pluto makes a long, slow trip around the Sun. It takes about 248 years for Pluto to orbit the Sun once.

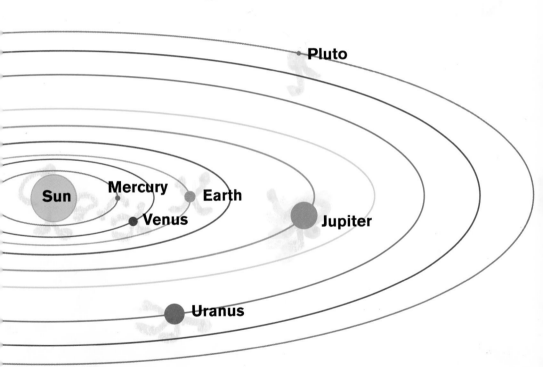

Pluto's path around the Sun is shaped like an oval. That means Pluto's distance from the Sun changes as it orbits.

Pluto is usually the farthest planet from the Sun. But sometimes Pluto and Neptune cross paths. Then Neptune becomes the farthest planet from the Sun. Pluto becomes the second farthest.

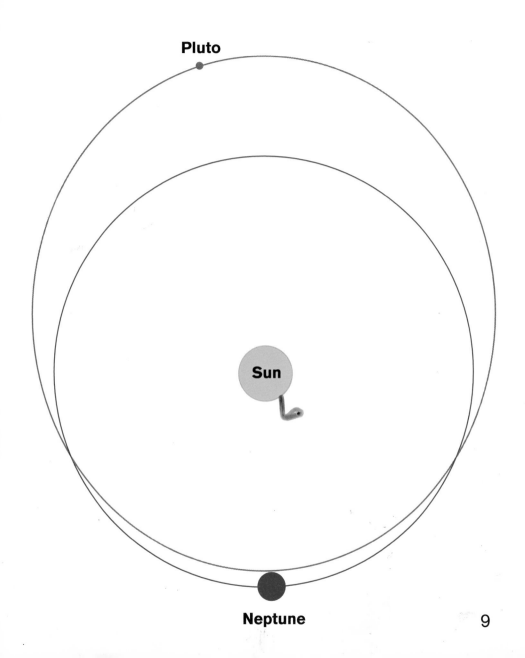

Pluto moves in another way besides orbiting. It spins around like a top. This kind of spinning is called **rotating.**

The other planets rotate, too. Earth takes one day to spin around once. Pluto takes about six days to rotate.

11

Pluto is a cold, icy world. It is always much colder than the coldest places on Earth. People have called it a space snowball. It is a rocky ball covered with ice. Does the ice ever melt?

Some of the ice melts when Pluto is closest to the Sun. The melted ice turns into a layer of gas. This gas is called the **atmosphere.** It freezes back into ice when Pluto travels away from the Sun.

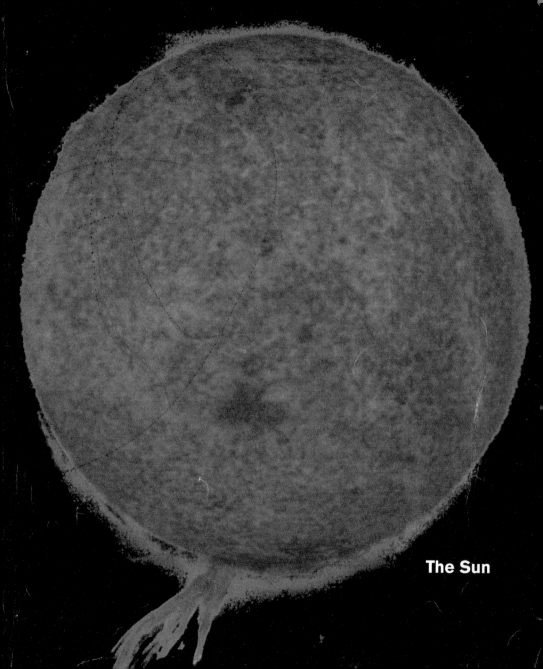

The Sun

Pluto has a close neighbor in space. It is a moon named Charon. A moon is a smaller body that orbits a planet.

Most moons are much smaller than the planets they orbit. But Charon is about half the size of Pluto. Some people think Charon is more like a planet than a moon. They call Pluto and Charon a double planet.

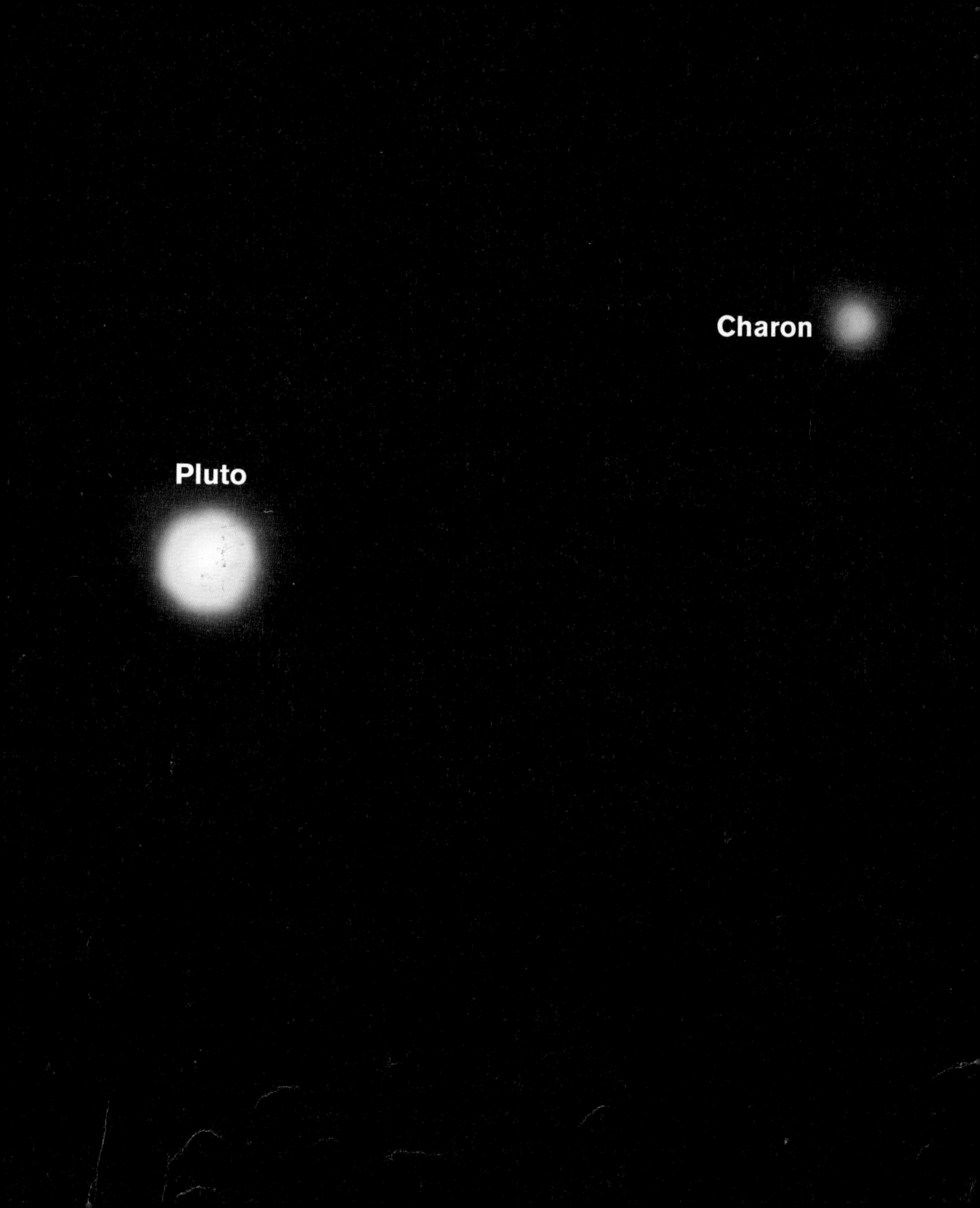
Charon
Pluto

People have watched the night skies for thousands of years. They could see some of the planets in the solar system. But no one knew about Pluto for many, many years.

Telescopes helped people learn more about our solar system. Telescopes make faraway objects look larger and closer. The first telescopes were invented in the 1600s.

Percival Lowell was the first person to search for Pluto. He was an **astronomer.** Astronomers are people who study outer space.

Lowell began looking for Pluto in 1905. He used special telescopes with cameras attached to them. The telescopes took pictures of the sky. Lowell searched for more than ten years. But he never found Pluto.

An astronomer named Clyde Tombaugh led the next search for Pluto. Tombaugh made photographs using a powerful new telescope. Then he studied the photographs carefully.

Tombaugh looked for anything that moved from one place to another in the pictures. In 1930, he made an important discovery. He saw that a bright dot had moved. The dot was Pluto.

In 1978, astronomers looked at some pictures made by a big telescope in Arizona. They discovered the moon Charon.

An artist made this picture of Charon orbiting Pluto.

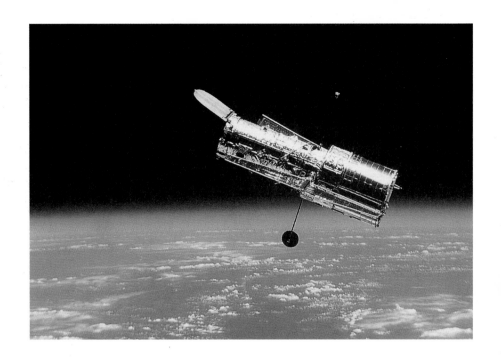

The Hubble Space Telescope began
taking photographs of the solar system
in the 1990s. It is the best telescope
ever made. But Pluto is too far away
for the Hubble to see it clearly.

No spacecraft has ever visited Pluto. A space mission to Pluto is planned for the future. A spacecraft will fly to the planet to take pictures and make tests. It will also study Charon.

The spacecraft will make a long trip. It will travel for at least eight years to get to Pluto.

This painting shows a spacecraft visiting
Pluto and Charon.

What will the spacecraft find when it gets to Pluto? What does the planet look like up close?

Imagine taking your own trip to Pluto. What do you think you would find?

An artist made this picture of Pluto, Charon, and the distant Sun

Facts about Pluto

- Pluto is 3,670,000,000 miles (5,900,000,000 km) from the Sun.

- Pluto's diameter (distance across) is 1,430 miles (2,300 km).

- Pluto orbits the Sun in 248 years.

- Pluto rotates in 6 days.

- The average temperature on Pluto is $-390°F$ $(-233°C)$.

- Pluto's atmosphere is made of methane and nitrogen.

- Pluto has one moon.

- Pluto was discovered in 1930 by Clyde Tombaugh.

- Pluto was named after the Roman god of the underworld.

- The first name for Pluto was Planet X.

- Pluto cannot be seen without a telescope.

- Pluto is 39 times farther from the Sun than Earth is.

- Pluto was the last of the nine planets in the solar system to be discovered.

Glossary

astronomer: a person who studies outer space

atmosphere: the layer of gases that surrounds a planet or moon

orbit: to travel around a larger body in space

rotating: spinning around in space

solar system: the Sun and the planets, moons, and other objects that travel around it

Learn More about Pluto

Books

Brimner, Larry Dane. *Pluto*. New York: Children's Press, 1999.

Wetterer, Margaret K. *Clyde Tombaugh and the Search for Planet X*. Minneapolis: Carolrhoda Books, Inc., 1996.

Websites

Solar System Exploration: Pluto
<http://solarsystem.nasa.gov/features/planets/pluto/pluto.html>
Detailed information from the National Aeronautics and Space Administration (NASA) about Pluto, with good links to other helpful websites.

The Space Place
<http://spaceplace.jpl.nasa.gov>
An astronomy website for kids developed by NASA's Jet Propulsion Laboratory.

StarChild
<http://starchild.gsfc.nasa.gov/docs/StarChild/StarChild.html>
An online learning center for young astronomers, sponsored by NASA.

Index